ARMADILLOS

by Martha London

Cody Koala
An Imprint of Pop!
popbooksonline.com

abdobooks.com
Published by Pop!, a division of ABDO, PO Box 398166, Minneapolis, Minnesota 55439. Copyright © 2021 by POP, LLC. International copyrights reserved in all countries. No part of this book may be reproduced in any form without written permission from the publisher. Pop!™ is a trademark and logo of POP, LLC.

Printed in the United States of America, North Mankato, Minnesota
082020
012021

THIS BOOK CONTAINS RECYCLED MATERIALS

Cover Photo: Shutterstock Images
Interior Photos: Shutterstock Images, 1, 5 (bottom left), 9 (top), 12-13, 16; iStockphoto, 5 (top), 5 (bottom right), 9 (bottom left), 9 (bottom right); Ignacio Yufera/Biosphoto/Science Source, 6; Mark Payne-Gill/Nature Picture Library/Alamy, 10; James Davies/Alamy, 15; Bianca Lavies/National Geographic, 19, 20

Editors: Christine Ha and Brienna Rossiter
Series Designer: Sophie Geister-Jones

Library of Congress Control Number: 2019954983
Publisher's Cataloging-in-Publication Data
Names: London, Martha, author.
Title: Armadillos / by Martha London.
Description: Minneapolis, Minnesota : POP!, 2021 | Series: Underground animals | Includes online resources and index.
Identifiers: ISBN 9781532167591 (lib. bdg.) | ISBN 9781532168697 (ebook)
Subjects: LCSH: Armadillos--Juvenile literature. | Armored animals--Juvenile literature. | Burrowing animals--Juvenile literature. | Underground areas--Juvenile literature.
Classification: DDC 599.3/12--dc23

Hello! My name is
Cody Koala

Pop open this book and you'll find QR codes like this one, loaded with information, so you can learn even more!

Scan this code* and others like it while you read, or visit the website below to make this book pop.

popbooksonline.com/armadillos

*Scanning QR codes requires a web-enabled smart device with a QR code reader app and a camera.

Table of Contents

Chapter 1
Safe in the Sand 4

Chapter 2
Hard Armor 8

Chapter 3
Insects for Dinner 14

Chapter 4
Growing Up 18

Making Connections 22
Glossary. 23
Index 24
Online Resources 24

Chapter 1

Safe in the Sand

Armadillos are **mammals**. They live in South America, Central America, and North America. They dig **burrows** in dry, sandy places.

Watch a video here!

5

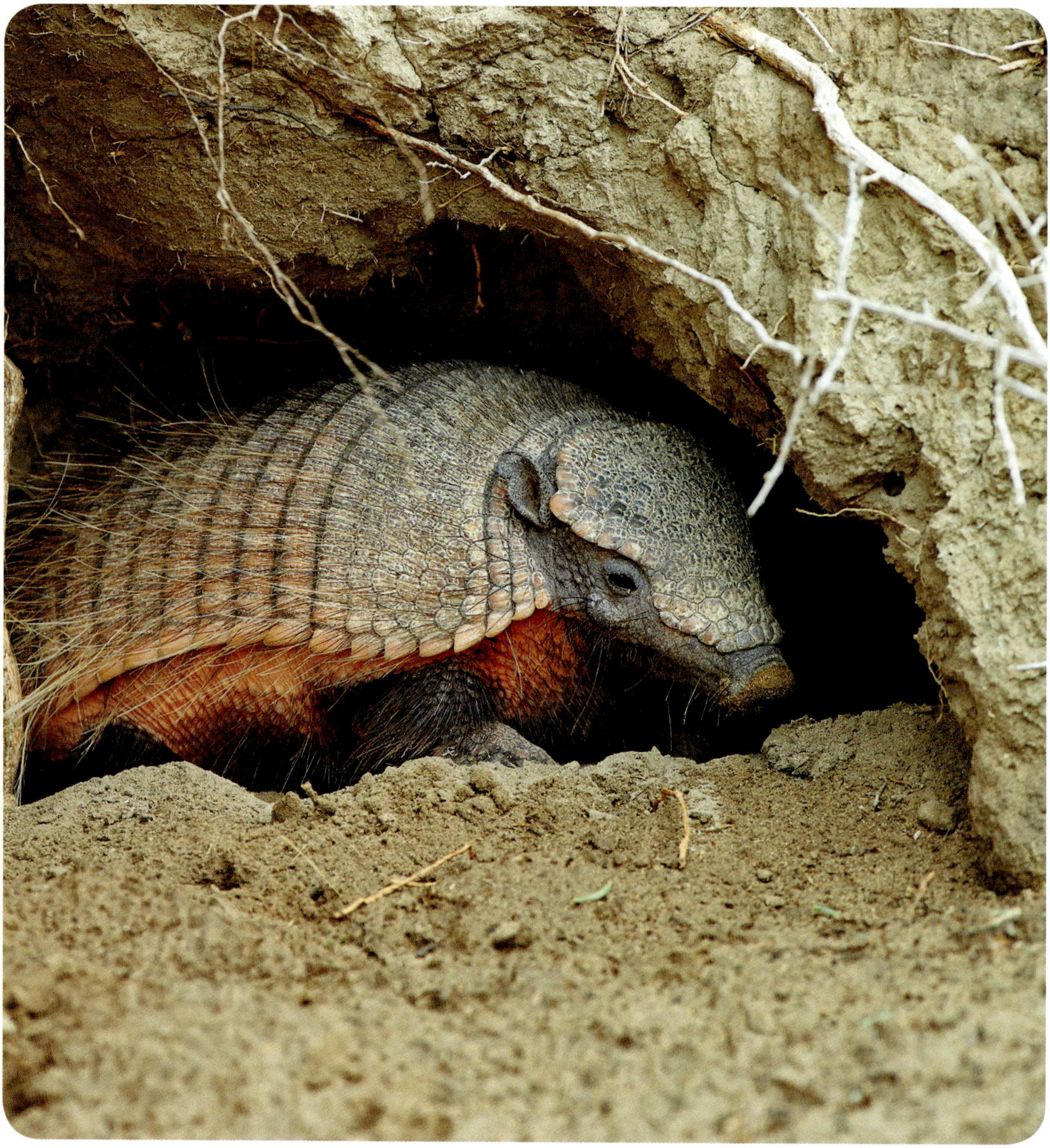

Armadillos use their burrows to hide from **predators**. Armadillos sleep in their burrows too. In cold weather, they use leaves and grass to make nests.

> The smallest armadillos are only 6 inches (15 cm) long. The biggest can reach 5 feet (1.5 m).

Chapter 2

Hard Armor

Armor covers an armadillo's back. The armor is made of hard plates. Some types of armadillos have just three plates. Others have up to 11.

Learn more here!

9

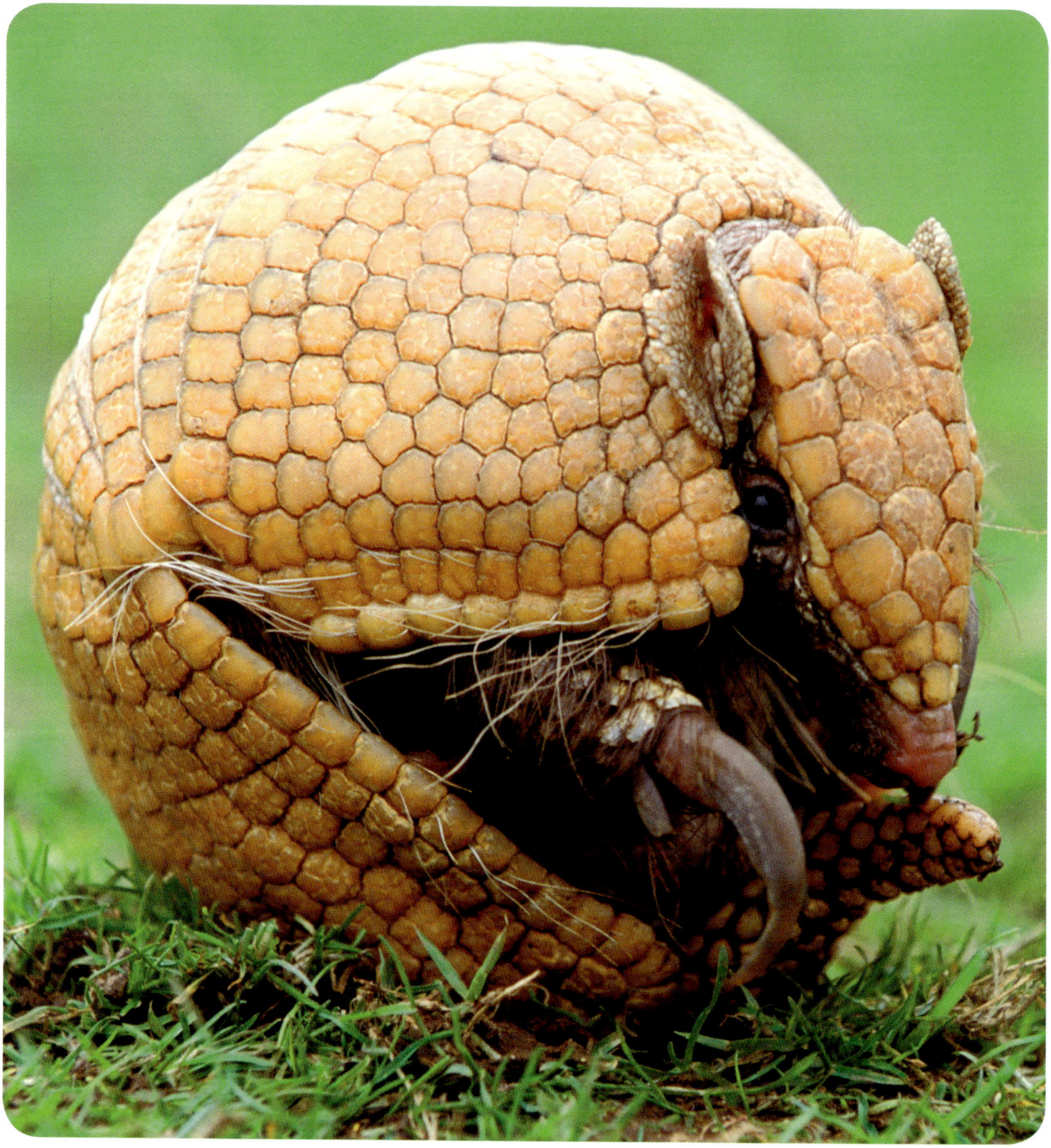

The plates help armadillos stay safe. Armadillos have soft bellies. If they sense danger, armadillos lie down flat. That way, only the hard plates show. Some armadillos can curl up. The plates act like a hard shell.

Armadillo means "little armored one" in Spanish.

armor

tail

Armadillos have sharp claws. The claws help armadillos dig. Armadillos

ear

head

claw

also have long, sticky tongues. Their tongues help them grab food.

Chapter 3

Insects for Dinner

Armadillos eat mostly insects. They eat some fruits and plants too. They often look for food in the morning and evening. They sleep during the day.

> Armadillos sleep up to 16 hours a day.

Learn more here!

15

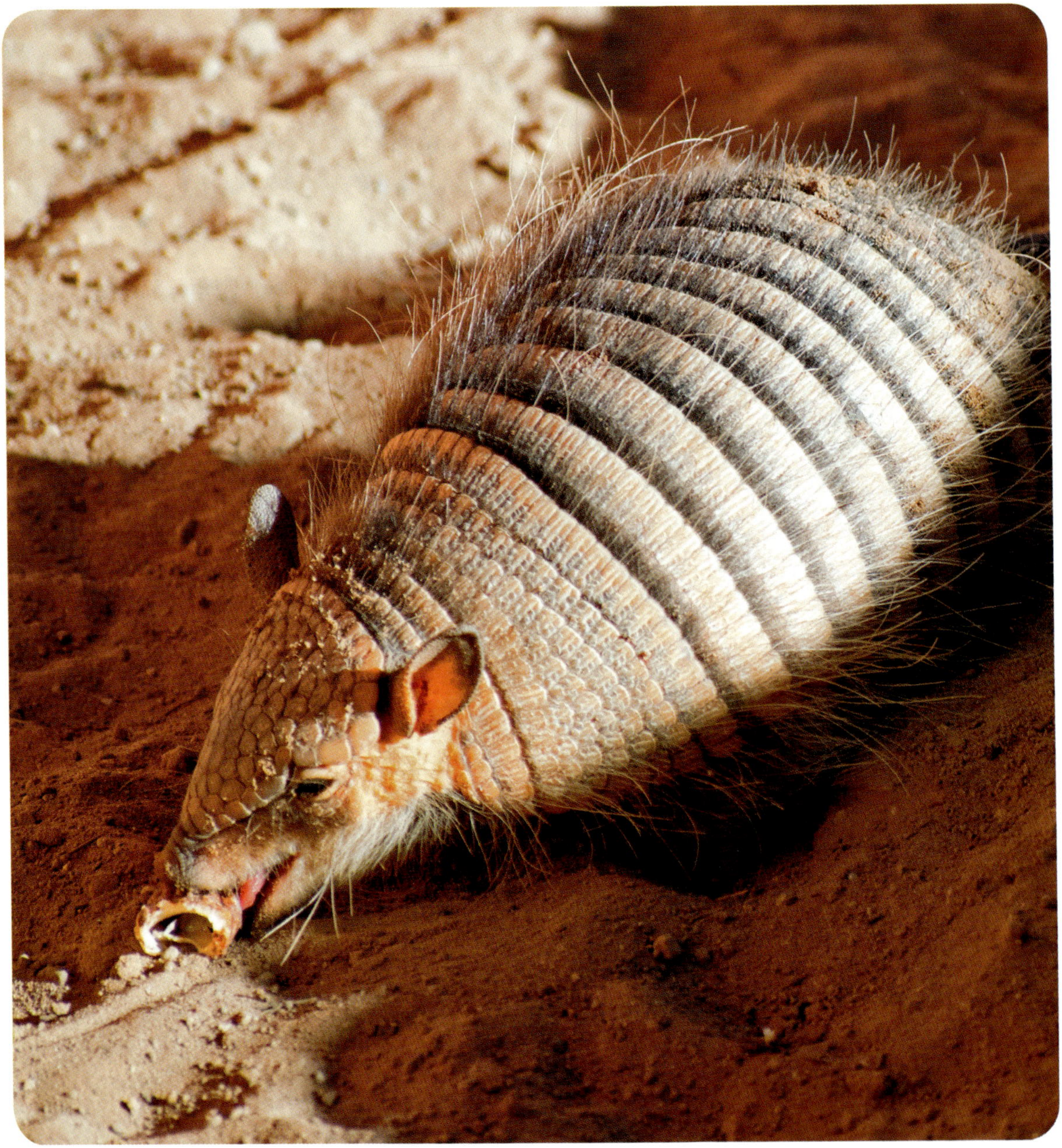

An armadillo uses its sense of smell to find food. The armadillo digs holes in the ground. Or it pulls apart tree bark. Then its tongue reaches inside to grab insects.

Chapter 4

Growing Up

Like most other **mammals**, armadillos have live babies. Females make nests in their **burrows**. They give birth to up to 12 **pups** at once.

Complete an activity here!

20

The pups drink milk. Their mother teaches them how to find food. Pups leave the burrow when they are four months old. Most full-grown armadillos live alone. In the wild, armadillos can live up to 30 years.

Making Connections

Text-to-Self

Would the area you live in be a good home for an armadillo? Why or why not?

Text-to-Text

What books have you read about other mammals? How are armadillos similar to those animals? How are they different?

Text-to-World

An armadillo's long, sticky tongue helps it grab insects. Can you think of other body parts that help animals catch food?

Glossary

armor – a hard outer covering that protects the body.

burrow – a hole that an animal digs in the ground for shelter.

mammal – a type of animal that has hair or fur and feeds milk to its young.

predator – an animal that hunts other animals for food.

pup – a baby animal.

Index

armor, 8, 12

burrows, 4, 7, 18, 21

claws, 12, 13

mammals, 4, 18

plates, 8, 11

predators, 7

pups, 18, 21

tongues, 13, 17

Online Resources

popbooksonline.com

Thanks for reading this Cody Koala book!

Scan this code* and others like it in this book, or visit the website below to make this book pop!

popbooksonline.com/armadillos

*Scanning QR codes requires a web-enabled smart device with a QR code reader app and a camera.